Co-Enzyme Q-10

Woodland Publishing
Pleasant Grove, UT

Woodland Publishing
P.O. Box 160
Pleasant Grove, UT
84062

Co-Enzyme Q-10

Introduction

Co Q-10 is a "co-enzyme" which is found in the cells of all breathing, living things. This nutrient is a catalyst which cells need for energy. With energy, cells reproduce properly and live longer and healthier. When our cells are in top shape, we feel much better. When we are young, our bodies and immune systems are in good condition. However, as we age our ability to synthesize Co Q-10 is greatly decreased, and that is when health troubles seem to begin.

According to researchers Emile G. Blinzakov, M.D. and Gerald L. Hunt, Co Q-10 can "boost the immune system, increase the strength of the heart without exercise, relieve angina, protect against heart attacks, lower high blood pressure, reduce weight naturally, heal periodontal problems, extend life, and slow down the aging process." They explain, "Co Q is a vital catalyst in the creation of the energy that cells need for life. Without CoQ the chain of cellular energy is broken, and without energy, life ceases." In *Whole Foods* magazine, (April 1978) Richard A. Passwater, Ph. D. discusses the role of nutrients in the body, and explains how Co Q-10 works: "Co Q-10 resembles vitamin E and vitamin K in chemical structure. Not too surprisingly, Co Q-10 biochemically functions much like vitamin E in that it participates in anti-oxidant and free-radical reactions. However, Co Q-10 has a special biochemical role of major importance; it is involved in the production of energy cells. Of special import is Co Q-10's role in producing energy in heart cells.

Vitamins often form the major part of coenzymes in the body. Pyridoxine (B-6), thiamin (B-1), riboflavin (B-2), niacin amide (B-3), pantothenic acid (B-5) and folic acid are examples of vitamins that are easily converted into coenzymes. Coenzymes should not be confused with enzymes. Coenzymes are relatively small molecules compared to enzymes, and thus are not inactivated by stomach acids and are readily absorbed. Enzymes are large proteins that have very specific roles in carrying out desirable chemical reactions. Enzymes are catalysts that speed reactions that might take place very slowly or not even take place at all if it were not for the presence of the enzymes.

Yet, the enzymes are not consumed in the reaction...each enzyme usually aids only one type of reaction. Many enzymes work only in the presence of a smaller organic molecule, the coenzyme.

All live foods contain one form or another of coenzyme Q-1 to coenzymes Q-10. This compound is so "ubiquitous" (found in many places or foods) that it was first named ubiquinone. Humans require CoQ. We can make CoQ out of other coenzymes (Q-1 to coenzyme Q-9), or we can absorb it from our food. However, we cannot manufacture CoQ from simple nutrients or CoQ's building blocks. As we age, many of us lose our ability to efficiently make CoQ our of the other coenzyme Q's (Q-1 to Q-9). Thus, many persons will depend on their food for all of their CoQ, and thus it becomes a vitamin to those individuals. To make matters worse, foods lose CoQ with processing and storage."

WHAT ARE SOME BENEFITS OF CO Q-10?

Co Q-10 demonstrates significant power when used by cardiac patients. The heart muscle has a special requirement for this co-enzyme. It is one of its most necessary nutrients. Take it away, and a myriad of cardiovascular problems ensue. Supply it in adequate amounts an the problems are reversed. In clinically controlled tests with heart patients it has been shown that angina pectoris responds favorably to Co Q-10. Likewise, when volunteers, who did not exercise or participate in similar fitness routines, took 60 mg. of Co Q-10 daily for two months, there was a surge in their ability to withstand vigorous exercise! Also researchers have discovered that when cardiac drugs and Co Q-10 are used simultaneously, the dosages of the heart medications can be reduced, yet their effectiveness is greatly enhanced.

One elderly man relates that he was on heart medications and hated to take the drugs but was told by his doctor that he had no choice...he had to take them the rest of his life. He was introduced to Co Q-10 and began taking it. Immediately he noticed he did not tire as easily as before. He discovered a surge of rejuvenation and energy. He decided to experiment. He reduced the amount of heart medicine he was taking to see if he would still have the same amount of energy. He still felt just as good, so reduced the medication further. (He was still taking the Co Q-10, and finally tapered off to the point where he ceased taking the heart pills altogether.) He took Co Q-10 exclusively from then on and still feels "on top of the world." This man is in his 80's! However, we are not recommending this. Any change in medication should be approached with extreme caution. (It is interesting to note that heart disease occurs around middle age, and it is around this same period of time that the body begins to slow down its production of Co Q-10.)

Co Q-10 can be of intense benefit to the immune system. Our immune systems are constantly bombarded with foreign micro-organisms. Twenty-four hours a day our bodies are fighting off molds, fungi, bacteria, viruses, toxins, etc. This is an exhaustive job. Sometimes when the stress load is too high the red warning light appears. This often comes in the form of sickness or chronic tired feeling. Co Q-10 has the unique ability to increase oxygen utilization and also activate the "germ-eaters" in our body. It is like the commanding officer marshaling our forces. When the immune system says, "No More!" and the immune 'soldiers' want to call it quits, Colonel Co Q gives them a shot of oxygen, which revives them. They then attack the foreign invaders with renewed vigor.

In research experiments conducted with mice at the New England Institute, Co Q-10 effected a longer life span for these little rodents. The average mouse lived 150 weeks when given supplemental Co Q-10. This equates to approximately 140 human years!

Co Q-10 is of significant value for periodontal, or "gum" disease. Dr. Edward G. Wilkinson, periodontal specialist and a pioneer in studies with Co Q-10 and this disorder, reports one of many case histories. In one instance, an elderly lady could not eat solid foods for months due to excruciating pain experienced every time she tried to chew anything. She was given Co Q-10 for two weeks, after which the pain subsided. She began eating regular food again and three months after that her gums had returned to their pink, healthy condition. This woman had been given no other treatment besides Co Q-10!

Startling new research has linked obesity with Co Q-10 deficiencies. Slim people who eat properly, and do not diet, have been given blood tests which showed higher amounts of Co Q-10 in their systems than overweight individuals. When a person diets, many times there is a subsequent lack of necessary vitamins, minerals, enzymes and coenzymes. In this way the obese person is "whipping a tired horse." The problem of obesity is only made worse by depriving the cells of nutrient like Co Q-10. A study conducted between the University of Texas and the University of Antwerp, Belgium, has demonstrated that weight loss can be achieved with Co Q-10 supplementation.

Dr. Lars Ernster, from the University of Stockholm states: "Ubiquinone (Co Q-10) exerts effects against free radicals as an antioxidant, similar to vitamin E. In fact, there is evidence for a protective effect of ubiquinone against lipid peroxidation that results in serious membrane damage and is also responsible for 'age granules' in aging cells." This means that Co Q-10 inhibits destruction of cells by free radicals by binding them with oxygen molecule and rendering them harmless. This action protects cells from damage. Cells in a healthy condition will contribute to youthfulness, well-being and a strong immune system. When cells are weakened, the signs of age and altered immunity becomes visible. According to one report a woman in her late thirties was disturbed with the apparent aging of her skin, as well as extreme fatigue. She heard about Co Q-10 and began taking it. Within one month she noticed a difference. Her energy level increased immensely and her skin had a more youthful, taut appearance.

Japanese studies with Co Q-10 have shown that this nutrient is beneficial for allergies and asthma. It has been demonstrated that Co Q-10 is an effective antihistamine. More over...sniffly, sneezy, allergy drugs...Co Q-10 has arrived on scene!

Dr. Karl Folkers, pioneer of Co Q-10 research explains its benefit for muscle disorders. He reports, " Over six years ago, an adult with late onset form of muscle disease contacted me to volunteer to join our research program as a patient, because he had been advised by his neurologist that the should mentally prepare himself for a wheelchair within two years. Eight months or so after this medical prediction, which devastated the morale of this lawyer, he was placed on Co Q-10.

I have long turned to Dr. Couch for medical advice, and when he reviewed this case, he remarked that if this patient could be kept out of a wheelchair for five years, we might as well take a greater interest in Co Q-10 and muscular dystrophy. This patient has now been on Co Q-10 for only a few days less then six years...not just five years. He not only is not in a wheelchair, by he swims, bowls, and plays golf frequently each week and conducts a vigorous business life in his legal profession.

Other diseases targeted for help using Co Q-10 include: multiple sclerosis, Alzheimer's disease, lupus, schizophrenia, asthma and allergies.

MORE RESEARCH

Research by Dr. Bliznokov and Gerald Hunt, conducted at the New England Institute uncovered still another remarkable effect of Co Q-10. They explain, "Further confirmation of the beneficial effect of Co Q-10 on the immune system was displayed in our experiments, which revealed its ability to reduce the number and size of chemically induced tumors in mice. It also increased the number of survivors.

The tumors were purposely induced by injection of dibenzpyrene, a chemical with very potent carcinogenic (cancer-causing) effect. As an illustration of this chemical's dangerous potency, it is commonly present in tar and is a prime suspect for producing lung tumors in smokers.

This type of chemically induced tumor is considered by scientists most closely resemble natural occurring tumors in humans. By the 55th day after the administrations of the carcinogen, 85 percent of the control (not treated with CoQ) group had developed tumors, and by day 69 it had extended to 100 percent. Meanwhile, among the CoQ treated mice only 25 percent displayed tumor growth at day 55; 65 percent at day 69, and only 77 percent at day 300!

Health author Jonathan Rothschild, in an article found in the 10-87 issue of *Let's Live!*, states: "Tissues and cells involved in immune function are highly energy dependent and therefore require sufficient supplies of Co Q-10. Numerous studies have documented an immune-enhancing effect of Co Q-10, including an improvement in the immunological profile of cancer patients. Co Q-10 supplementation my help prevent or reverse age-related immuno-suppression...Psychiatrists who routinely prescribe antidepressant and antipsychotic medications are concerned about the cardiotoxicity of these compounds. In two clinical studies, Co Q-10 supplementation improved EKG (electrocardiogram) changes in patients on psychotropic drugs and reduced the incidence of drug-induced arrhythmias."

Dr. Robert M. Downs, D.C., when interviewed by Alice Van Baak for *Bestways* magazine had more to add on the benefits demonstrated by Co Q-10 supplementation.

He states: "It has been used experimentally in patients with congestive heart failure. One special report was on seven patients with mild congestive heart failure who received 30 milligrams of Co Q-10 a day. All patients showed improvement, and 53 percent showed no more symptoms."

Dr. Downs explained further: "I frequently make nutritional recommendations or patients undergoing chemotherapy and radiation therapy...to keep their immune systems functioning as correctly as possible, and to maintain their weight. Obviously, I'm interested in doing everything possible to help prevent alteration in their white and red blood cell counts. I reviewed an article on the use of Co Q-10 in helping to prevent some of the cardiotoxicity associated with the use of medication.

This particular report involved 11 cancer patients being treated with Adriamycin. The study stated that their Co Q-10 levels were significantly lower than in the control groups. It was also reported that this particular medication inhibits Co Q-10 dependent enzymes, and was probably the main cause of Co Q-10 deficiency. In the test animals of this report, Co Q-10 was given to seven patients, beginning three to five days before the drug treatment. An additional seven patients received Adrimycin without Co Q-10. It was a small group, but the study suggested that Co Q-10 may prevent (or help prevent) cardiotoxicity in Adriamycin-treated patients.

If this is true, it would allow desperate cancer patients to tolerate an effective chemotherapeutic agent, instead of having to withdraw because of side effects. Another series of reports that I studied presented some evidence that Co Q-10 deficiency has been found in cardiac and skeletal muscles of animals with hereditary muscular dystrophy...I frequently have difficulty accepting animal research as being the same as human research and therapy. However, it was intriguing to note the deficiencies of Co Q-10 that have been found in muscles of humans with muscular dystrophy. In one double-blind study, 100 milligrams of Co Q-10 was given daily for three months to 12 patients with progressive muscular dystrophy. The report related that Co Q-10 treatments resulted in significant improvement of cardiac output and stroke volume, as well as increased physical well-being in four out of the eight subjects.

Subjective improvement (according to the report) included reduced leg pain, better control of leg function, less fatigue and increased exercise tolerance. I think a great deal of research is needed in this specific area, because if a small percentage of individuals with this devastating condition could be helped by a naturally occurring component, it should be deeply investigated..."

CONCLUSION

Co Q-10 is naturally occurring nutrient in foods. However, it is destroyed by modern processing methods. Even fresh fruits and vegetables, if they were picked days ago and allowed to sit around, have diminished quantities of nutrients...vitamins, minerals, and Co Q-10.

Since most foods contain very little Co Q-10, and our ability to synthesize it diminishes with age, additional supplementation can ensure a sufficient supply. Everyone can benefit from Co Q-10.

It may well be the answer to the "fountain of youth", as well as health!